Planet Earth

Discover and understand our world's natural wonders

ENCYCLOPÆDIA

Britannica®

CHICAGO   LONDON   NEW DELHI   PARIS   SEOUL   SYDNEY   TAIPEI   TOKYO

PROJECT TEAM

Judith West, *Editorial Project Manager*
Christopher Eaton, *Editor and Educational Consultant*
Indu Ramchandani, *Project Editor (Encyclopædia
    Britannica India)*
Bhavana Nair, *Managing Editor (India)*
Rashi Jain, *Senior Editor (India)*
Kathryn Harper, *U.K. Editorial Consultant*
Colin Murphy, *Editor*
Locke Petersheim, *Editor*
Nancy Donohue Canfield, *Creative Director*
Megan Newton-Abrams, *Designer*
Amy Ning, *Illustrator*
Joseph Taylor, *Illustrator*
Karen Koblik, *Senior Photo Editor*
Paul Cranmer, *Retrieval Specialist and Indexer*
Barbara Whitney, *Copy Supervisor*
Laura R. Gabler, *Copy Editor*
Dennis Skord, *Copy Editor*
Marilyn L. Barton, *Senior Production Coordinator*

**ENCYCLOPÆDIA BRITANNICA
PROJECT SUPPORT TEAM**

EDITORIAL

Theodore Pappas, *Executive Editor*
Lisa Braucher, *Data Editor*
Robert Curley, *Senior Editor, Sciences*
Brian Duignan, *Senior Editor, Philosophy, Law*
Laura J. Kozitka, *Senior Editor, Art, World Culture*
Kathleen Kuiper, *Senior Editor, Music, Literature,
    World Culture*
Kenneth Pletcher, *Senior Editor, Geography*
Jeffrey Wallenfeldt, *Senior Editor, Geography, Social Sciences*
Anita Wolff, *Senior Editor, General Studies*
Charles Cegielski, *Associate Editor, Astronomy*
Mark Domke, *Associate Editor, Biology*
Michael Frassetto, *Associate Editor, Religion*
James Hennelly, *Associate Editor, Film, Sports*
William L. Hosch, *Associate Editor, Math, Technology*
Michael R. Hynes, *Associate Editor, Geography*
Michael I. Levy, *Associate Editor, Politics, Geography*
Tom Michael, *Associate Editor, Geography*
Sarah Forbes Orwig, *Associate Editor, Social Sciences*
Christine Sullivan, *Associate Editor, Sports, Pastimes*
Erin M. Loos, *Associate Editor, Human Biology*
Anne Eilis Healey, *Assistant Editor, Art, World Culture*

DESIGN

Steven N. Kapusta, *Designer*
Cate Nichols, *Designer*

ART

Kathy Nakamura, *Manager*
Kristine A. Strom, *Media Editor*

ILLUSTRATION

David Alexovich, *Manager*
Jerry A. Kraus, *Illustrator*

MEDIA ASSET MANAGEMENT

Jeannine Deubel, *Manager*
Kimberly L. Cleary, *Supervisor, Illustration Control*
Kurt Heintz, *Media Production Technician*
Quanah Humphreys, *Media Production Technician*

CARTOGRAPHY

Paul Breding, *Cartographer*

COPY

Sylvia Wallace, *Director*
Larry Kowalski, *Copy Editor*
Carol Gaines, *Typesetter*

INFORMATION MANAGEMENT/INDEXING

Carmen-Maria Hetrea, *Director*

EDITORIAL LIBRARY

Henry Bolzon, *Head Librarian*
Lars Mahinske, *Geography Curator*
Angela Brown, *Library Assistant*

EDITORIAL TECHNOLOGIES

Steven Bosco, *Director*
Gavin Chiu, *Software Engineer*
Bruce Walters, *Technical Support Coordinator*
Mark Wiechec, *Senior Systems Engineer*

COMPOSITION TECHNOLOGY

Mel Stagner, *Director*

MANUFACTURING

Dennis Flaherty, *Director*

INTERNATIONAL BUSINESS

Leah Mansoor, *Vice President, International Operations*
Isabella Saccà, *Director, International Business Development*

MARKETING

Patti Ginnis, *Senior Vice President, Sales and Marketing*
Jason Nitschke, *National Sales Manager, Retail Advertising
    and Syndication*
Michael Ross, *Consultant*

ENCYCLOPÆDIA BRITANNICA, INC.

Jacob E. Safra,
Chairman of the Board

Ilan Yeshua,
Chief Executive Officer

Jorge Cauz,
President

Dale H. Hoiberg,
Senior Vice President and Editor

Marsha Mackenzie,
Managing Editor and Director of Production

Planet Earth

INTRODUCTION

What's an oasis? Is a "finger of land" smaller than your hand?
How can water be stronger than stone?
What's another name for giant floating ice cubes?

In *Planet Earth*, you'll discover answers to these questions and many more. Through pictures, articles, and fun facts, you'll travel around the world, seeing the highest and the lowest, the hottest and the coldest, and the strangest and most beautiful places on Earth.

To help you on your journey, we've provided the following guideposts in *Planet Earth*:

■ **Subject Tabs** — The colored box in the upper corner of each right-hand page will quickly tell you the article subject.

■ **Search Lights** — Try these mini-quizzes before and after you read the article and see how much — *and how quickly* — you can learn. You can even make this a game with a reading partner. (Answers are upside down at the bottom of one of the pages.)

■ **Did You Know?** — Check out these fun facts about the article subject. With these surprising "factoids," you can entertain your friends, impress your teachers, and amaze your parents.

■ **Picture Captions** — Read the captions that go with the photos. They provide useful information about the article subject.

■ **Vocabulary** — New or difficult words are in **bold type**. You'll find them explained in the Glossary at the end of the book.

■ **Learn More!** — Follow these pointers to related articles in the book. These articles are listed in the Table of Contents and appear on the Subject Tabs.

Britannica
LEARNING LIBRARY

Have a great trip!

The pyramid and the camel, pictured here in Giza, Egypt, are two images often associated with the Egyptian desert.

Planet Earth

TABLE OF CONTENTS

Learning About the Earth

Geography is a science that studies the Earth's surface. It studies what makes the different shapes and colors of the Earth—the ground, rocks, and water, what does and does not grow.

If you look at the Earth as a geographer does, then you might see it as a colorful map. Much more than half of it is blue with oceans, lakes, rivers, and streams. In some places it is tan-colored with the sands of dry deserts. In other places it is green with forests. There are purple-gray mountains and white snowcapped peaks. And there are the soft yellow of grainfields and the light green of leafy crops.

Part of learning about the Earth is learning where people can and can't live. The different colors of your Earth map can help you discover this.

You won't find many people in the tan, white, or larger blue parts— deserts, the snowfields, and oceans. Not many people live in the deserts, because deserts are hot and dry. Very few plants can grow there. In the high mountains and at the North and South poles, it is very cold. Most plants don't like the cold, and most people don't either.

You will find people in and near the green and yellow parts and the smaller blue parts—the farmlands, forests, rivers and lakes. To those regions you can add brown dots and clusters of dots, for towns and cities.

There's a lot to learn about the Earth, just as there's a lot to learn about a friend. Geography helps you become a friend of the Earth.

LEARN MORE! READ THESE ARTICLES...
CONTINENTS • OCEANS • RAINFORESTS

The coast of Nova Scotia, in Canada, shows some of the Earth's many shapes and colors. Geography looks closely at what makes these different shapes and colors.
© Raymond Gehman/Corbis

DID YOU KNOW?
The "big blue marble" is a nickname for the planet Earth. This is because from space our world looks like a big round marble, all blue with swirling white streaks of clouds.

Areas where not many people live are also the areas where few plants grow. Why do you think that is? (Hint: What do you do with lettuce, beans, and apples?)

Answer: If few plants grow in an area, then few animals will live there. This is because animals need either plants or other animals to eat. And without plants or animals, there's nothing for people to eat.

The Largest Pieces of Land

NORTH AMERICA

The continents are the largest bodies of land on the Earth. Look at a globe. Whatever is blue is water. Most of the rest is land: the continents.

There are seven continents. From biggest to smallest, they are Asia, Africa, North America, South America, Antarctica, Europe, and Australia.

Some continents, such as Australia and Antarctica, are completely surrounded by water. And some continents are joined together, as Asia and Europe are.

Continents are physical bodies, defined by their shape, size, and location. They have mountains, rivers, deserts, forests, and other physical features. But humans have divided them into **political** groups, called "countries" or "nations."

Large continents, such as Asia, may include both very large countries, such as China, and very small countries, such as Nepal. Australia, the smallest continent, is also itself a country—one of the world's largest.

North America contains three large countries—Canada, the United States, and Mexico—and a few small countries in a region known as Central America. Europe, on the other hand, is the world's second smallest continent but has about 50 countries.

Africa, the second largest continent, is believed to be where the very first humans appeared. The continent of Antarctica is all by itself down at the South Pole. It is rocky and is covered by thick ice that never melts. Only a few plants and animals can be found along its seacoasts.

Earth scientists believe that the continents began forming billions of years ago. Lighter parts of Earth's **molten** core separated from heavier parts and rose to the top. As they cooled off and became solid, the land that would become the continents formed.

The continents were probably joined together at first and then drifted apart. One theory supposes that there were once two "supercontinents": Gondwanaland in the south and Laurasia in the north.

LEARN MORE! READ THESE ARTICLES...
ANTARCTICA • DESERTS • OCEANS

SOUTH AMERICA

SEARCH LIGHT

Name the seven continents.

8

EUROPE

ASIA

AFRICA

AUSTRALIA

DID YOU KNOW?

Here's a silly rhyme to help you remember the continents:

Africa is hot,
Antarctica is cold.
Asia is crowded,
Europe is old.
There's an America down South,
and one up North too,
And Australia has the kangaroo.

ANTARCTICA

Answer: Africa, Antarctica, Asia, Australia, North America, South America, and Europe.

A Continent of Extremes

Antarctica is the coldest, windiest, and highest **continent** in the world! It lies at the bottom of the world, surrounding the South Pole. The name Antarctica means "opposite to the Arctic," referring to the Arctic Circle on the other side of the world.

The coldest temperature recorded in Antarctica is also the world's lowest, at –128.6° F. A sheet of ice covers the entire continent. At its thickest point, the ice is almost 3 miles deep—and that's on *top* of the ground. The continent contains most of the world's ice and much of the world's freshwater. Toward the edges of the continent, the ice becomes glaciers, creeping rivers of ice.

Strange and wonderful Antarctica has only one day in the entire year. The Sun generally rises on September 21 and sets on March 22. This one long day is the summer! From March 22 until September 21, the South Pole is dark and Antarctica has its night, or winter.

People do not live permanently in Antarctica. Only scientists and some adventurous tourists visit. There are, however, 45 species of birds in

DID YOU KNOW?
Antarctica is a desert—a "frigid desert." It's extremely cold, unlike the more common hot sandy deserts. But like them, it gets so little moisture during the year that very little life can survive.

Antarctica, including the emperor penguin and the Adélie penguin, that live near the seacoast. Also, four species of seals breed only in Antarctica.

Whales live in the water around the **frigid** continent. The killer whale, the sperm whale, the rare bottle-nosed whale, the pygmy whale, and seven species of baleen whales can all be found off the coast.

Oddly, there are active volcanoes in Antarctica. That means you can find not just the world's coldest temperatures here but, deep down, some of the hottest too.

LEARN MORE! READ THESE ARTICLES...
DESERTS · GLACIERS · ICEBERGS

These emperor penguins are some of Antarctica's very few inhabitants. So in a way they might indeed be considered the "rulers" of this harsh and beautiful frozen desert continent.
© Galen Rowell/Corbis

SEARCH LIGHT

Match the numbers with the correct labels. You may have to do some figuring and clever thinking!

−128.6 *bird species*
182.5 *thickness of ice*
3 *length of one day*
45 *coldest temperature*

Answer: 45 — *bird species*
3 (miles) — *thickness of ice*
182.5 (days) — *length of one day*
−128.6 (° F) — *coldest temperature*

11

Building Earth's Giant Landscapes

What makes mountains? Several different processes contribute to mountain building. And most mountains are formed by a combination of these, usually over millions of years.

Deep inside, the Earth is so incredibly hot that everything is melted, or molten. This molten material, or lava, escapes to the Earth's surface when volcanoes erupt. The lava cools and becomes hard and solid. This happens again and again, collecting until there is a volcanic mountain.

Mount Fuji in Japan and Mount St. Helens in Washington state, U.S., are volcanic mountains. There are also many undersea volcanic mountains—much taller than anything on land!

In some cases strong earthquakes caused the surface rock for miles and miles to break. Part of the surface would then be lower and part of it higher. More earthquakes moved the lower parts down and the upper parts up. Eventually, the high parts became tall enough to make mountains.

Still other mountains were pushed up from the bottom of an ocean when two enormous portions of the Earth crashed together—*very slowly*, over millions and millions of years. Some of the largest mountain chains formed this way. The Andes of South America are an example.

Another mountain-building process is called "folding." If you push a carpet up against a wall, it folds and rumples. That's basically the way the Appalachian Mountains in eastern North America were formed.

At first most mountains were steep and sharp. But even hard rocks can be worn away. Slowly, with the wind and the rain rubbing at them, steep sharp mountains grow smoother, shorter, and rounder.

LEARN MORE! READ THESE ARTICLES…
CAVES • ISLANDS • OCEANS

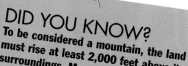

DID YOU KNOW?
To be considered a mountain, the land must rise at least 2,000 feet above its surroundings. Mount Everest, the world's highest mountain, rises 29,035 feet above sea level.

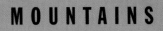

DID YOU KNOW?

Here's a good rhyming way to remember which formation is a stalactite and which is a stalagmite. Stalactites hold "tight" to the ceiling. Stalagmites "might" reach the roof.

When Water Is Stronger than Stone

Caves are natural openings in the Earth large enough for a person to get in. Most have been made when rainwater or streams have worn away rock—usually a softer rock such as limestone. The wearing-away process is called "erosion."

Slowly, over millions of years, the water works away at the soft rock, making a small tunnel-like opening. As more and more rock wears away, the opening grows wider and deeper. Soon even more water can flow in. In time, many of these openings become huge caves, or caverns.

Mammoth Cave-Flint Ridge in Kentucky is a linked system of caverns. It is 345 miles long, one of the longest in the world. In France the Jean Bernard, though much shorter (11 miles long), is one of the world's deepest caves, reaching down more than 5,000 feet.

Some caves have beautiful craggy formations called "stalactites," like those pictured here, that hang from the cave's roof. These are made by water seeping into the cave. Each drop leaves a very tiny bit of dissolved rock on the ceiling of the cave. After thousands and thousands of years, an icicle-shaped stalactite forms.

When water drips to the cave's floor, it deposits small **particles** of solids. These slowly build up into a stalagmite, which looks like an upside-down icicle.

There are other kinds of caves that are made in different ways. When lava flows out of a volcano, it sometimes leaves gaps, making volcanic caves. When ice melts inside a glacier, glacier caves result. And ocean waves pounding on the shore year after year can wear away a cave in the face of a cliff.

LEARN MORE! READ THESE ARTICLES...
GLACIERS • GRAND CANYON • RIVERS

SEARCH LIGHT

Which of the following is *not* a way that caves are formed?
ocean waves
lava
lightning
water erosion
ice melts

Answer: Caves aren't formed by lightning.

Lands of Little Water

Deserts are places that get very little rain each year—so little rain that most trees and plants cannot grow there. Some deserts will go for years without rain. They are difficult places to live in, and the few plants,

Golden desert snapdragons, or yellow Mojave flowers, in Death Valley, California, U.S.
© Darrell Gulin/Corbis

animals, and people who live there have to be tough to survive. Every continent except Europe has a desert. Even Antarctica has one, a **frigid** desert.

Most deserts, however, are arid, or dry, deserts with mile after mile of sand, baked earth, and barren rock. In the daytime these places look like lost worlds—hot, dry, and silent. Usually, the only plants growing there are low thorny ones. These plants store most of the water they are able to collect. It may be a long time before their next drink.

At night it can be quite cold in the desert. That's when creatures that have been hiding from the Sun's burning rays come out of their homes. Many of the creatures are lizards and insects such as scorpions. There are also different kinds of rats as well as other, larger animals.

You can hear the animals squeaking and growling near water holes and springs. That's where the coyotes, badgers, bobcats, foxes, and birds gather, all hunting for food and water. When the rare spring does bubble up in the desert, plants and trees begin to grow. An island of green like this is called an "oasis."

Many people choose to live in the desert. In late afternoon the sky turns crimson and gold, and the mountains make purple shadows. And at night the stars seem close enough to touch.

LEARN MORE! READ THESE ARTICLES...
ANTARCTICA • OASIS • RAINFORESTS

SEARCH LIGHT

Fill in the blank: Every continent except

_____

has a desert.

This California (U.S.) desert, called Death Valley, is both beautiful and dangerous. It's also the lowest point below sea level in the Western Hemisphere.
Joseph Sohm—Chromosohm/Photo Researchers

DID YOU KNOW?

Desert sands are known to "sing." For some reason that scientists do not yet fully understand, sand sometimes makes a booming, barking, or humming noise when walked upon or moved by some other natural force.

Answer: Every continent except **Europe** has a desert.

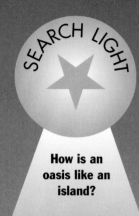

SEARCH LIGHT

How is an oasis like an island?

In the Sahara desert an oasis like this depends heavily on date palms. They provide both food and enough shade to grow other plants that are too sensitive to grow directly in the desert sun.
Robert Everts–Stone/Getty Images

Water in the Desert

Probably the most precious thing in the world is fresh water. If a person was lost in a desert without any special equipment or supplies, he or she would soon die from lack of water.

It is therefore not surprising that very few people live in the desert. But some people do. Where do they stay? Obviously, they stay where there is water.

A place in the desert with a natural supply of fresh water is called an "oasis." An oasis has enough water to support a variety of plants.

Most oases (the plural of "oasis") have underground water sources such as springs or wells. Al-Hasa is the largest oasis in the Middle Eastern country of Saudi Arabia. It has acres and acres of palm groves and other crops.

But not all oases have a constant supply of water. Some areas have dry channels called "wadis," where springs sometimes flow. And desert areas at higher elevations sometimes receive extra rain to support plant life.

In the Sahara people can live year-round in the oases because the water supply is permanent. The oases allow crops to be watered, and desert temperatures make crops grow quickly. The date palm is the main source of food. However, in its shade grow citrus fruits, figs, peaches, apricots, vegetables, and cereals such as wheat, barley, and millet.

The Siwa Oasis in western Egypt has about 200 springs. It is a very fertile oasis, and thousands of date palms and olive trees grow there. In fact, the people living in this oasis export dates and olive oil to other places in the world.

LEARN MORE! READ THESE ARTICLES…
DESERTS · ISLANDS · NILE RIVER

DID YOU KNOW?
Few people realize just how extreme desert weather can be. The hottest desert temperature recorded is 136° F, in Libya. And in Chile there is a desert that apparently hasn't had any rain for the last 400 years.

Answer: An oasis is like an island of water surrounded by a sea of sand. It's kind of a reverse island.

Fingers of Land

A peninsula is a body of land surrounded by water on three sides. The word "peninsula" comes from the Latin *paene insula*, meaning "almost an island." There are peninsulas on every **continent**, but every one is different. Most peninsulas of any significance extend into oceans or very large lakes.

In the United States, Florida is a peninsula. The state of Alaska qualifies as one and has several smaller peninsulas of its own.

One of the last great wilderness areas in the United States is on the Olympic Peninsula in Washington state. It is surrounded by the Pacific Ocean, the Strait of Juan de Fuca, and Puget Sound. It has a rainforest, rivers, **alpine** peaks, glaciers, and such creatures as salmon and elk.

In Mexico there are two main peninsulas, the Yucatán Peninsula in the east and Baja California in the west. The Yucatán Peninsula draws tourists to the ruins of great Mayan cities such as Uxmal and Chichén Itzá.

Another famous peninsula is the Sinai Peninsula of Egypt. It is triangular in shape. The peninsula links Africa and Asia. In Jewish history the Sinai Peninsula is known as the site where God appeared before Moses and gave him the Ten Commandments.

Europe too has several peninsulas. In northern Europe the Scandinavian Peninsula contains the countries of Norway and Sweden. Denmark forms another. And the Iberian Peninsula in southern Europe is made up of Spain and Portugal. Italy and part of Greece are peninsulas as well.

The world's largest peninsula is Arabia, at over a million square miles. Other important peninsulas in Asia include Korea and Southeast Asia.

LEARN MORE! READ THESE ARTICLES…
CONTINENTS · ISLANDS · OCEANS

DID YOU KNOW?

Peninsulas such as Iberia (Spain and Portugal), Italy, and Florida tend to be popular tourist destinations. For example, Florida gets almost 59 million tourists a year.

SEARCH LIGHT

Which of the following are peninsulas? (Feel free to consult your classroom map or globe.)

Korea   Britain
Portugal   Arabia
Italy   Denmark
Hawaii   Florida

Answer: The only two that are *not* peninsulas are Hawaii and Britain. They are islands.

Endangered Ecosystems

Imagine a forest with a carpet of wet leaves littering the ground. If you look up, you see only a **canopy** of broad green leaves. There are wildflowers on the trees. You can hear water drops, insects, birds, and, perhaps, the distant screech of a monkey. The place you are picturing is a rainforest.

A rainforest is a kind of **ecosystem**—a community of all the living things in a region, their physical environment, and all their interrelationships.

Rainforests are dense, wet, and green because they get large amounts of rain. The Amazon Rainforest in South America is the world's largest

View of the Venezuelan rainforest canopy from the air.
© Fotografia, Inc./Corbis

rainforest. Other large rainforests lie in Central Africa and Southeast Asia. Northeastern Australia's "dry rainforest" has a long dry season followed by a season of heavy rainfall.

In a rainforest nothing is wasted. Everything is **recycled**. When leaves fall, flowers wilt, or animals die on the forest floor, they decay. This releases nutrients into the soil that become food for the roots of trees and plants. Water **evaporates** in the forest and forms clouds above the trees. Later this water falls again as rain.

Rainforests are rich in plants and animals. Many have not even been discovered yet. Some rainforest plants have given us important medicines. These include aspirin, which is a pain reliever, and curare, used to help people relax during medical operations.

Unfortunately, the rainforests are being destroyed rapidly. The trees are felled for **timber** and to create land for farming. Animals living in these forests are facing extinction. And once lost, these animals and forests cannot be replaced.

LEARN MORE! READ THESE ARTICLES…
AMAZON • DESERTS • OASIS

DID YOU KNOW?

Rainforests are being cut down or burned at an alarming rate. Scientists estimate that every day a rainforest the size of New York City is lost.

SEARCH LIGHT

What's one important way that rainforests help people? (Hint: Think of aspirin.)

Answer: Rainforest plants have helped unlock the secrets of many of the drugs we use to keep ourselves healthy today. Aspirin is one of these.

SEARCH LIGHT

What's one way that swamps and marshes are alike? What's one way that they're different?

Grassy Wetlands

A marsh is a wetland, an area of land containing much soil moisture that does not drain well. Swamps are also wetlands. The main difference is that while trees grow in a swamp, grasses grow in a marsh. Marsh grasses have shallow roots that spread and bind mud together. This slows the flow of water, which creates rich soil deposits and encourages the growth of the marsh.

DID YOU KNOW?
The largest marsh in the world is the Florida Everglades. This marsh-swamp combination is somewhat more than 2,000 square miles and is home to many extraordinary animals, including the very rare Florida panther.

There are two main types of marshes, freshwater marshes and salt marshes. Freshwater marshes are found at the mouths of rivers. These marshes are famous as bird **sanctuaries** and are an important **habitat** for many birds, mammals, and insects. If we didn't have the marshes, then we would lose many of these animals. There simply isn't anywhere else where they can survive.

The Amazon in South America, the Congo in Africa, the Nile in Egypt, the Tigris and Euphrates in Iraq, and the Mekong in Vietnam all have large freshwater marshes.

Did you know that the rice you eat grows in freshwater marshes? Rice is the most important of all marsh plants. It provides a major portion of the world's food.

Salt marshes are formed by seawater flooding and draining flat land as tides go in and out. The grasses of a salt marsh will not grow if the ground is permanently flooded. Salt marshes are found along the east coast of the United States, in the Arctic, in northern Europe, in Australia, and in New Zealand.

LEARN MORE! READ THESE ARTICLES…
RAINFORESTS • RIVERS • TIDES

The Ruby Marshes in the state of Nevada, U.S., provide a great example of what these grassy wetlands look like.
© David Muench/Corbis

Answer: Both swamps and marshes are wetlands and support a lot of wildlife. But while trees grow in swamps, grasses grow in marshes.

The Power
of Flowing Water

It seems pretty obvious what rivers are for. They give us water to drink and fish to eat. They do these things for many animals too. But it might surprise you to learn that rivers have some even bigger jobs.

For one thing, rivers deliver water to lakes and oceans. Another major task is changing the face of the land, and this second job makes a huge difference. No other force changes as much of the world's surface as running water does. In fact, the world's rivers could completely **erode** the face of the Earth, though it might take them 25 million years to do it.

We can see rivers' **handiwork** all around us. Valleys are carved out when rivers slowly cut through rock and carry off dirt. Canyons and gorges are young valleys.

SEARCH LIGHT

Fill in the blank: You could describe one of a river's main jobs as being a sculptor of _____.

Another impressive bit of river handiwork is the waterfall. Waterfalls happen when a river wears away soft rock and then drops down onto hard rock that it can't erode. Some falls are **harnessed** to produce electricity.

The world's tallest waterfall is Angel Falls in Venezuela. It drops an incredible 3,212 feet. Khone Falls on the Mekong River in Southeast Asia sends 2 1/2 million gallons of water over the edge every second—the most of any falls and nearly double the flow of North America's Niagara Falls.

The world's longest river is the Nile in North Africa. The Amazon in South America is a little shorter but carries more water than any other river.

LEARN MORE! READ THESE ARTICLES…
FLOODS • GRAND CANYON • NILE RIVER

Answer: You could describe one of a river's main jobs as being a sculptor of land.

27

Engulfed by Water

Take a small bowl and place a sponge in it. Now slowly pour water into the bowl. The sponge soaks up the water. But once the sponge is full, the bowl begins to fill up with water. If you pour more water, the bowl will overflow.

This is what happens in a flood. The ground is like a giant sponge that soaks up rainwater until it is full. Some of the water dries and goes back into the air. The rest, called "runoff," can't be soaked up and can cause floods.

There are different types of floods. Spring floods occur when heavy winter snows melt rapidly. Floods caused by heavy rains can occur at any time of the year. Rivers overflow their banks, and the ground can't soak up the extra water.

The rain and wind accompanying hurricanes (or typhoons, in the Pacific Ocean) can also cause floods. Huge ocean waves **overwhelm** coastal towns, and the heavy rains cause rivers and streams to flood nearby areas. Such hurricane-created floods struck Central America in 1998, killing more than 20,000 people and leaving one and a half million homeless.

A flash flood, however, comes without warning. When a **cloudburst** occurs in hilly country or in a dry riverbed, the runoff is fast. The ground doesn't have time to soak up the rainwater. Destructive flash floods happen when a great deal of water overflows all at once.

Volcanic eruptions and earthquakes at sea may cause huge waves, called "tsunamis," that may swamp seacoasts. The volcanic eruption of Krakatoa in 1883 formed waves that flooded whole districts in Indonesia.

LEARN MORE! READ THESE ARTICLES…
NILE RIVER • RIVERS • TIDES

In 1999 these people and others suffered losses in the
floods that followed Hurricane Irene in Florida, U.S.
© AFP/Corbis

SEARCH LIGHT

**Fill in
the blanks:
When it rains,
the _____
soaks up the water.
Water that doesn't get
soaked up is called
"_____."**

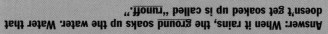

**Answer: When it rains, the ground soaks up the water. Water that
doesn't get soaked up is called "runoff."**

29

DID YOU KNOW?
In 1998 Christian Taillefer of France set a cycling speed record. He rode down the face of a glacier on a bicycle and reached a speed of 132 miles an hour.

Rivers of Ice

In high mountains there are places that are packed full of ice. These ice packs are called "glaciers" and look like giant frozen rivers. And like rivers, glaciers flow—but usually so slowly you can't see them move.

It takes a long time to make a glacier. First, snow falls on the mountains. It collects year after year, until there is a thick layer called a "snowfield."

In summer the surface of the snowfield melts and sinks into the snow below it. There it freezes and forms a layer of ice. This too happens year after year, until most of the snowfield has been changed into ice. The snowfield is now a glacier.

The snow and ice in a glacier can become very thick and heavy. The glacier then begins to actually move under its own weight and creeps down the mountain valley. It has now become a valley glacier.

SEARCH LIGHT

A valley glacier is
a) a glacier that has grown up in a valley.
b) a thick layer of snow.
c) a glacier that has started to move down a mountain.

The valley glacier moves slowly but with enormous force. As it moves, it scrapes the sides of the mountain and tears off pieces of it. Sometimes it tears off chunks as big as a house. As the glacier moves down the mountain into warmer regions, the ice begins to melt. The icy water fills rivers and streams.

Many thousands of years ago, much of the Earth's surface was covered with moving glaciers. This period is sometimes called the Ice Age. As the world warmed, most of the ice melted away and formed many of the rivers, lakes, and seas around us today—including the Great Lakes in North America, which have an area greater than the entire United Kingdom.

LEARN MORE! READ THESE ARTICLES…
ICEBERGS • MOUNTAINS • RIVERS

In Alaska's Glacier Bay National Park, the 16 glaciers that descend from the mountains present an amazing sight.
© Neil Rabinowitz/Corbis

Answer: c) a glacier that has started to move down a mountain.

The Rainforest River

On a map of South America a thick line cuts across the country of Brazil all the way from the Andes Mountains in Peru to the Atlantic Ocean. That line traces the mighty Amazon River. The other lines that lead into it are major rivers too. Altogether they make up one of the world's greatest river systems, carrying more total water than any other.

Why is the river called "Amazon"? Many years ago, in 1541, a Spanish soldier named Orellana sailed down the river. He had to fight many women soldiers who lived by the river. It made him think of the Amazons, who were the mighty women soldiers of Greek mythology. So he called the river "Amazon."

Along the banks of the river are miles of trees, all tangled together with bushes and vines. This region is known as the "rainforest." You can hear the sound of water dripping from leaves because it rains here almost every day. This is the largest tropical rainforest in the world.

In the rainforest there are very tall trees, some as tall as 200 feet. They spread out like giant umbrellas and catch most of the sunlight. There are rubber trees, silk cotton trees, Brazil nut trees, and many others. Many animals, some quite rare, make their homes among the tree branches. These include **exotic** parrots and **mischievous** monkeys—as well as giant hairy spiders!

LEARN MORE! READ THESE ARTICLES...
NILE RIVER • RAINFORESTS • RIVERS

SEARCH LIGHT

True or false? It rains almost every day in the Amazon.

ATLANTIC OCEAN

DID YOU KNOW?

You may have heard "Amazon" used for a totally different thing: Amazon.com. Perhaps this Internet store wanted to be the greatest of its kind, just as the Amazon River is.

Amazon River

Peru

Brazil

SOUTH AMERICA

ATLANTIC OCEAN

Answer: TRUE. The Amazon region is a very wet kind of area known as a "rainforest."

Egypt's Gift

There's one country that depends almost entirely on the river that flows through it. That country is Egypt, and the river is the Nile.

Life would be **drastically** different in Egypt if there was no Nile River. The river is the source of all the water used for farming in Egypt. That is why people call Egypt the "gift of the Nile."

People farm on the banks of the river. Two of the most important things they grow are rice and cotton. Egyptian cotton is one of the finest cottons in the world.

It rains very little in Egypt. Where it does, it's not much more than seven inches a year. There are very dry deserts on both sides of the Nile River. The plants you will find there are mostly thorny bushes and desert grass.

Long ago even Egypt's seasons depended on the river. There were just three seasons. *Akhet* was when the river was flooded. During *peret* the land could be seen after the flood. And *shomu* took place when the river's waters were low.

The Nile River is **teeming** with different kinds of fish. The most common is the Nile perch. The river is also an important waterway. Canals, or man-made streams, act as a highway **network** for small boats and ships during the flood season.

After its long journey across North Africa, the Nile empties into the Mediterranean Sea.

LEARN MORE! READ THESE ARTICLES...
AMAZON • FLOODS • RIVERS

SEARCH LIGHT

Which of the following descriptions matches the term *shomu*?
a) gift of the Nile
b) the flooding season
c) the low-water season
d) the season after the flood

DID YOU KNOW?

The Aswan High Dam was built across the Nile in the 1960s and '70s. Amazingly, the entire ancient Egyptian temple center of Abu Simbel was moved to keep it from being submerged.

Answer: c) the low-water season

"The Smoke That Thunders"

It is difficult to stand in front of this spectacular African waterfall without feeling small. Victoria Falls is about twice as high as Niagara Falls in North America and about one and a half times as wide. It inspires awe and respect in all who see it.

Victoria Falls lies on the border between Zambia and Zimbabwe in southern Africa along the course of the Zambezi River.

The falls span the entire **breadth** of the Zambezi River at one of its broadest points. There is a constant roaring sound as the river falls. A dense blanket of mist covers the entire area. The Kalolo-Lozi people who live in the area call this mist Mosi-oa-Tunya, "the Smoke That Thunders."

The first European to set eyes on this wonder of nature was the British explorer David Livingstone. He named it after Queen Victoria of the United Kingdom.

The waters of Victoria Falls drop down a deep **gorge**. All the water of the Zambezi River flows in through this gorge. At the end of the gorge is the Boiling Pot, a deep pool into which the waters churn and foam during floods. The river waters then emerge into an enormous zigzag channel that forms the beginning of the Batoka Gorge.

The Victoria Falls Bridge is used for all traffic between Zambia and Zimbabwe. When it opened in 1905, it was the highest bridge in the world.

In 1989 Victoria Falls and its parklands were named a World Heritage site.

LEARN MORE! READ THESE ARTICLES...
GEOGRAPHY • NIAGARA FALLS • RIVERS

© Patrick Ward/Corbis

DID YOU KNOW?
Victoria Falls is huge, but another waterfall takes the record for being the tallest. Angel Falls in Venezuela dwarfs Victoria at an amazing 3,212 feet tall.

SEARCH LIGHT

The average
height of
Niagara Falls
is about 165 feet.
What height would
you estimate for
Victoria Falls?
(Hint: Look in the first
paragraph.)

Answer: Victoria Falls is about twice as high as Niagara Falls. So you can estimate Victoria Falls is 330 feet high.

DID YOU KNOW?

The great Niagara Falls once stopped running. On the evening of March 29, 1848, ice blocked the Niagara River, and the falls stopped for about 30 hours.

Thunder of Waters

Niagara Falls, one of the most spectacular natural wonders in North America, is more than 25,000 years old. The falls are on the Niagara River, which flows between the United States and Canada, from Lake Erie to Lake Ontario.

Horseshoe Falls, the Canadian section of Niagara Falls.
© Dave G. Houser/Corbis

It is awesome just to be near the waterfall and watch the force of so many gallons of water plunging down the steep cliff. More than 600,000 gallons per second pour from the falls. As the water thunders down, it fills the air with a silvery mist in which you can see many shining rainbows. A ceaseless roar fills the air as all this water strikes the bottom. The Iroquois Indians called this waterfall Niagara, meaning "thunder of waters."

The falls are divided into two parts by Goat Island. The larger portion is the Canadian section, known as Horseshoe Falls. It measures 2,600 feet along its curve and drops 162 feet. The American Falls are smaller and rockier. Their width is 1,000 feet across, and they drop about 167 feet.

Between the American Falls and Goat Island are the small Luna Island and the small Luna Falls, also called Bridal Veil Falls. There are caves behind the curtain of water of both these falls. One of these is the Cave of the Winds.

The best views of the falls are from Queen Victoria Park on the Canadian side, Prospect Point on the U.S. side, and Rainbow Bridge, which crosses between the two.

LEARN MORE! READ THESE ARTICLES…
CAVES • RIVERS • VICTORIA FALLS

SEARCH LIGHT

Find and correct the error in the following sentence: Niagara Falls is more than 2,500 years old.

Answer: Niagara Falls is more than 25,000 years old.

Nature's Masterpiece

A canyon is a deep steep-walled valley cut through rock by a river. The word "canyon" comes from the Spanish word *cañón*, which means "tube." Such valleys are found where river currents are strong and swift. A smaller valley cut out in the same way is called a "gorge."

Rafting through the Grand Canyon on the Colorado River.
© Kevin Fleming/Corbis

The Grand Canyon, in northern Arizona in the United States, is the most beautiful and awesome canyon in the world. It is cut a mile deep into the earth and is known for its fantastic shapes and colors. Swiftly flowing water, sand, gravel, and mud gave the rocks their interesting shapes. Each of its rock layers has a different shade of color, including **buff**, gray, green, pink, brown, and violet.

The canyon is 277 miles long and has been carved through the Arizona landscape by the Colorado River. It stretches in a zigzag course from the northern end of Arizona to the Grand Wash Cliffs near Nevada.

Many ancient pueblos—Native American cliffside dwellings—and other ruins in the canyon show that prehistoric peoples lived there. The Grand Canyon was first discovered by Europeans in 1540. It was established as a national park in 1919.

Visitors to the park can take a mule-pack trip down to the bottom of the canyon. People can even go river rafting, taking a thrilling trip over the rapids.

If you visit the canyon, you'll probably see some of the many animals that live there. Squirrels, coyotes, foxes, deer, badgers, bobcats, rabbits, chipmunks, and kangaroo rats all make their homes near the canyon.

In 1979 the Grand Canyon was named a World Heritage site.

LEARN MORE! READ THESE ARTICLES...
CAVES • GEOGRAPHY • RIVERS

SEARCH LIGHT

Fill in the blank: The word "canyon" comes from the Spanish word for "_____."

The Colorado River, seen here in the Marble Canyon portion of the Grand Canyon, cut the whole canyon—over millions of years.
Gary Ladd

DID YOU KNOW?
In geologic terms the Grand Canyon is fairly young, at about 6 million years old. But the rocks it cuts through date back as far as 2 billion to 2.5 billion years ago.

Answer: The word "canyon" comes from the Spanish word for "tube."

41

SEARCH LIGHT

How does
the ocean
help plants
to grow?

The World of Water

Did you know that nearly three-fourths of the Earth's surface is underwater? And almost all of that water is in one of the four major oceans. From biggest to smallest the oceans are: the Pacific, the Atlantic, the Indian, and the Arctic. Seas, such as the Mediterranean and the Caribbean, are divisions of the oceans.

© Kennan Ward/Corbis

The oceans are in constant motion. The **gravity** of the Moon and the Sun pulls on the oceans, causing tides—the regular rising and falling of the ocean along beaches and coastlines. The Earth's **rotation** makes the oceans circulate clockwise in the Northern **Hemisphere** and **counterclockwise** in the Southern Hemisphere. And winds cause waves to ripple across the ocean surface, as well as helping currents to flow underneath.

Currents are like rivers within the ocean. Some are warm-water currents, which can affect temperatures on land, and some are cold-water currents, which generally flow deeper. Major ocean currents, such as the Gulf Stream off the North American coast, also make for faster ocean travel.

We know less about the oceans than we do about the Moon. The ocean depths hide dramatic deep trenches and enormous mountain ranges. The Mid-Oceanic Ridge is a 40,000-mile range that circles the globe.

Oceans affect our lives in important ways. They provide fish to eat. They add moisture to the air to form clouds. And the clouds then make the rain that helps plants grow. Some scientists are even working on affordable ways to turn salt water into fresh water for drinking, cooking, washing, and watering crops. If they succeed, it will be one of the most important inventions of our time.

LEARN MORE! READ THESE ARTICLES...
ATLANTIC OCEAN • PACIFIC OCEAN • TIDES

DID YOU KNOW?

The Mariana Trench near the island of Guam has the deepest spot measured so far, at nearly seven miles. The world's highest mountain, Mount Everest, could sink in that spot and still have a mile of water above it.

Answer: Ocean water helps plants grow by adding moisture to the air, which turns into clouds. When the clouds gather enough moisture, it rains, which helps plants grow.

Dry Spots in a Watery World

Islands are areas of land surrounded on all sides by water. Islands come in all shapes and sizes. The very smallest are too small to hold even a house. The largest islands contain whole countries.

If you live in England, Iceland, Australia, or Japan, you live on an island. But these islands are so large that you might walk all day and never see water.

How do islands develop in the first place?

Some islands begin as fiery volcanoes in the ocean. Hot

Small island in the South Pacific Ocean.
© Craig Tuttle/Corbis

lava pours out of the volcano, making the island bigger and bigger. Slowly, as the lava cools, it becomes solid land, and when it rises above the water, it becomes an island. These are the volcanic islands.

Other islands are actually parts of the world's **continents**. Some of the land toward the edge of the continent may have been worn away over many, many years by wind or rain, or perhaps some of it sank. Then water from the ocean filled the low places and made a new island.

A row of islands may once have been the tops of mountains in a mountain range. The Aleutian Islands off the coast of North America were probably once a part of a mountain range that connected Alaska with Asia.

Maybe most surprising are the islands that are built up from the bottom of the ocean from the skeletons of tiny sea animals called "coral." As some corals die, others live on top of them. After thousands of years a coral island rises to the ocean surface. And these islands go on living!

LEARN MORE! READ THESE ARTICLES...
GREAT BARRIER REEF • OASIS • PENINSULAS

SEARCH LIGHT

Find and correct the error in the following sentence: Coral islands are made of tiny ocean rocks that have piled on top of each other for thousands of years.

This photo from the air shows one of the islands of the Maldives, a country made up of about 1,300 islands in the Indian Ocean.
© Lawson Wood/Corbis

44

If you try to count the number of islands in the world by looking at a globe, you'll probably come up with 300 or so. But that's only the major islands. Altogether the total is closer to 130,000.

Answer: Coral islands are made of tiny ocean creatures [or creatures' skeletons] that have piled on top of each other for thousands of years.

The Islands at the End of the World

A tortoise as big as a bathtub!

Giant lizards that look like dragons!

These are only a few of the special things that make the Galapagos Islands different from any other place on Earth. The islands lie in the Pacific Ocean, far away from any other land. People have called them "the world's end." Together with other natural wonders, the Galapagos are a World Heritage site.

The Galapagos Islands were formed from volcanoes that erupted in the sea. The bare and rocky islands look as if no creature could ever live there. But thousands of animals do, including many found nowhere else in the world. One animal that lives there is the giant tortoise, or land turtle. In fact, the islands got their name from these tortoises. The word *galápagos* means "tortoises" in Spanish.

One of the many varieties of finches on the Galapagos Islands.
© Galen Rowell/Corbis

The Galapagos Islands were especially important to the famous English scientist Charles Darwin. When Darwin visited the islands, he discovered that there were creatures living there that did not live anywhere else in the world. He saw three-foot-long lizards—land iguanas that looked like small dragons. And he saw amazing **marine** iguanas, lizards that had actually learned to swim. He also found a great many birds called "finches" that were all much the same except for differences in their beaks. These differences meant that they all ate different things, which allowed them all to share the same habitat.

Darwin decided that all plants and animals evolve, or change little by little, as the world around them changes. One plant or animal group will usually be more successful than another. Darwin called this process "natural selection." And he called the overall change through time the "theory of evolution."

Do you think people are evolving? What do you think we might look like in a million years?

LEARN MORE! READ THESE ARTICLES…
GEOGRAPHY · ISLANDS · PACIFIC OCEAN

The giant Galapagos tortoise can live as long as 150 years—longer than almost any other animal. Sadly, few are left today.
© Craig Lovell/Corbis

SEARCH LIGHT

Darwin's famous theory is called
a) natural selection.
b) good versus evil.
c) the big bang.

DID YOU KNOW?
The Galapagos finches all developed from the same ancestor. But to share such a small area, different groups developed beaks suited to different feeding habits. This fact helped Darwin understand how species change.

Island of Reefs
Within Reefs

The Great **Barrier** Reef is one of the great natural wonders of the world. It is actually a system of many individual reefs and islets (small islands). Altogether there are 2,100 individual reefs in the Great Barrier Reef. This huge ridge of coral reefs is separated from land by a **channel** of water 30 miles wide.

The Great Barrier Reef was formed over millions of years from mounds of coral. A coral is a soft animal that supports its body inside a hard hollow shell. When the coral died, its shell remained, and other corals grew on top of it. Over those millions of years, the corals remained hardened and became cemented together. Slowly they were covered with underwater plants, **debris** from the ocean, and other corals.

Exploring the Great Barrier Reef at low tide.
© Staffan Widstrand/Corbis

The Great Barrier Reef lies in the Pacific Ocean off the coast of Queensland in northeastern Australia. From north to south its length is equal to the entire Pacific Coast of the United States, extending more than 1,250 miles! The water is so clear and pollution-free at the Great Barrier Reef that people can glimpse the wonderful **marine** life deep underwater.

Besides the 400 types of coral, there are such animals as anemones, snails, lobsters, prawns, jellyfish, giant clams, and dugongs. And there are more than 1,500 **species** of saltwater fish. Many of the small fish have brilliant colors and unusual shapes.

The Great Barrier Reef was named a World Heritage site in 1981.

LEARN MORE! READ THESE ARTICLES…
ISLANDS • OCEANS • PACIFIC OCEAN

SEARCH LIGHT

True or false? The Great Barrier Reef is made of rock.

From north to south, the length of Australia's Great Barrier Reef is equal to that of the entire Pacific Coast of the United States!
© Australian Picture Library/Corbis

DID YOU KNOW?

Though it isn't truly a single structure, the 135,000-square-mile Great Barrier Reef is often referred to as the largest structure ever built by living things. And people had no hand in it!

Answer: FALSE. It's made of coral skeletons and live coral.

The Biggest Ice Cubes

Icebergs are simply broken-off pieces of glaciers or polar ice sheets that float out into the ocean. Very big pieces. Even little icebergs called "growlers" are as big as a bus. Big ones are longer than a freight train and as high as a skyscraper.

One especially surprising thing about an iceberg is that the part you see above the water is only a small bit of the whole iceberg. Most of the iceberg is underwater. You can see the way an iceberg floats by doing an easy experiment at home.

Fill a clear glass half full of very cold water. Drop an ice cube into the glass. Notice how most of the ice cube stays below the water.

The ice cube floats just the way an iceberg floats. And as the cube melts, it turns over, just as an iceberg does. Icebergs melt when they float away from freezing waters into warmer waters. Icebergs always start in the part of the world where it stays cold all the time, near the North or the South Pole.

Icebergs can be very dangerous when they float, big and silent, into the path of a ship. In the past many ships were wrecked because they hit an iceberg. Fortunately, this hardly ever happens anymore. This is because most modern ships have radar that finds the icebergs before they become a problem.

In addition, special airplanes from the International Ice Patrol watch for icebergs in likely areas, and satellites scan the oceans every day. Maps and warnings are regularly sent by radio to all the ships in nearby waters.

LEARN MORE! READ THESE ARTICLES...
ANTARCTICA • GLACIERS • ISLANDS

You can see from the size of the boat how big some icebergs actually are. But the much larger part of an iceberg is under the water!
Pal Hermansen—Stone/Getty Images

SEARCH LIGHT

Icebergs
are broken-off
pieces of
a) islands.
b) glaciers.
c) ice cubes.

Answer: b) glaciers.

The Ocean's Rise and Fall

Perhaps you have been to the beach and put your towel really close to the water. Then, when it was time to leave, the water seemed to have shrunk and was now far away from your towel.

At low tide the water slips low down on the beach. At high tide it will creep back up.
© Tim Thompson/Corbis

What actually happens is even more surprising. At high tide the water creeps up the beach. At low tide the water slips down. So the water really doesn't shrink; it simply goes away! But how, and where?

Most seashores have about two high tides and two low tides per day. It takes a little more than 6 hours for the rising waters to reach high tide. It takes another 6 hours for the falling waters to reach low tide. This 12-hour rise and fall is called the "tidal cycle."

Tides are caused mainly by the gravity of the Moon and the Sun pulling on the Earth. This causes ocean waters to pile up in a big bump of water directly beneath the Sun and the Moon. As the Earth **rotates**, the tidal bumps try to follow the two heavenly bodies.

The Sun and the Moon are in line with the Earth during a full moon or a new moon. Their gravity added together causes higher-than-normal high tides called "spring tides." When the Moon and the Sun are farthest out of line, their gravity forces offset each other. This causes lower-than-normal high tides, called "neap tides."

The tides in the Bay of Fundy in Canada rise higher than 53 feet. Beach towels and umbrellas at the Bay of Fundy don't stand a chance!

LEARN MORE! READ THESE ARTICLES…
CONTINENTS • FLOODS • OCEANS

DID YOU KNOW?

Some narrow rivers that empty into the sea develop large waves when extremely high tides rush into them. These waves, called "tidal bores," force the river's flow to change direction as they pass.

© Tim Thompson/Corbis

52

At high tide the water creeps high up on the beach.

It takes 6 hours for the tide to rise or fall. When the tide has both risen and fallen, it equals one tidal cycle. How long does it take for two tidal cycles?

Answer: Each tidal cycle has one rising tide and one falling tide. It takes 6 hours for the tide to rise or fall, so it takes 12 hours for it to do both. That is, 12 hours for one tidal cycle. Two tidal cycles then take 24 hours.

True or
false?
The Atlantic
is the saltiest
ocean.

SOUTH AMERICA

DID YOU KNOW?

Legend says that the Atlantic Ocean hides the remains of Atlantis, an island that supposedly sank beneath the sea. People have believed stories of Atlantis for many hundreds of years and have spent almost as much time searching for it.

EUROPE

The Youngest Ocean

The Atlantic Ocean is the world's second largest ocean, after the Pacific. It covers nearly 20 percent of the Earth. If you tasted water from all the oceans, you'd find the Atlantic to be the saltiest. And even though it is very old, it is actually the youngest ocean.

The Atlantic Ocean lies between Europe and Africa on one side of the globe and North and South America on the other. It reaches from the Arctic Ocean in the north to Antarctica in the south.

Like all oceans, the Atlantic has large movements of water **circulating** in it called "currents." Atlantic water currents move **clockwise** in the northern half of the world, but **counterclockwise** in the southern half. The Gulf Stream, a powerful and warm current in the North Atlantic, moves along the east coast of North America. There and elsewhere, the Gulf Stream has important effects on the weather.

Millions of tons of fish are caught each year in the waters of the Atlantic Ocean. In fact, more than half of all the fish caught in the world come from the Atlantic. The Atlantic is also used for activities such as sailing, windsurfing, and whale watching.

But despite the usefulness and magnificence of the Atlantic Ocean, the level of pollution has increased. People have allowed fertilizers, pesticides, and waste from toilets and sinks and factories to get into the ocean waters. As people and businesses try harder to stop pollution, the Atlantic will again become a healthier home for its animal and plant life.

LEARN MORE! READ THESE ARTICLES...
MEDITERRANEAN SEA · OCEANS · PACIFIC OCEAN

AFRICA

Answer: TRUE.

Arabian Peninsula

AFRICA

India

SEARCH LIGHT

Fill in
the blank
with the
correct number:
The Indian Ocean
is _____
times as big as the
United States.

DID YOU KNOW?
The world's longest mountain chain
is the undersea Mid-Ocean Ridge. It
stretches from the Arctic Ocean
through the Atlantic and Indian oceans
to the Pacific Ocean. The ridge is four
times as long as the Andes, Rockies,
and Himalayas *combined!*

ANTARCTICA

56

Ocean Between Many Continents

Millions of years ago, there was one huge mass of land in the Southern **Hemisphere**. It was the continent of Gondwanaland. But over many, many years Gondwanaland slowly broke up into the continents of South America, Africa, Antarctica, and Australia, as well as most of India.

The water that filled the growing space between these continents is now the Indian Ocean. The Indian Ocean is a huge body of salt water. It is the third largest ocean in the world—about five and a half times the size of the United States!

People from India, Egypt, and ancient Phoenicia (now mostly in Lebanon) were the first to explore this ocean. Later, Arabian merchants set up trade routes to the east coast of Africa. And Indian traders and priests carried their civilization into the East Indies. The dependable winds from the rainy season known as the "monsoon" made these voyages possible.

Today the Indian Ocean has major sea routes. They connect the Middle East, Africa, and East Asia with Europe and the Americas. Ships carry tanks of **crude oil** from the oil-rich Persian Gulf and Indonesia. The oil is important to modern society, but spills from these oil tankers can endanger ocean life.

The Indian Ocean is alive with plants, as well as animals such as sponges, crabs, brittle stars, flying fish, dolphins, tuna, sharks, sea turtles, and sea snakes. Albatross, frigate birds, and several kinds of penguins also make their home there.

AUSTRALIA

LEARN MORE! READ THESE ARTICLES...
ATLANTIC OCEAN · OCEANS · PACIFIC OCEAN

Answer: The Indian Ocean is 5½ times as big as the United States.

ATLANTIC OCEAN

The Sea in the Middle of Land

Italy

The Mediterranean Sea gets its name from two Latin words: *medius*, meaning "middle," and *terra*, meaning "land." The Mediterranean Sea is almost entirely surrounded by land. It's right between Africa, Europe, and Asia.

The Mediterranean is a bit larger than the African country of Algeria. But more important than its size is its location. Its central position made the Mediterranean an important waterway for a number of classical cultures, such as those of Italy, Greece, Egypt, and Turkey.

Many **channels** connect the Mediterranean with other bodies of water. The Strait of Gibraltar connects the Mediterranean with the Atlantic Ocean. The Dardanelles and the Bosporus connect it with the Black Sea, between Europe and Asia. And the Suez Canal is a man-made channel connecting the Mediterranean Sea with the Red Sea, which lies between the Arabian **Peninsula** and North Africa.

AFRICA

Three major rivers also lead into the Mediterranean Sea: the Rhône in France, the Po in Italy, and the Nile in Egypt. But the water from most of the rivers **evaporates** very fast. Instead, the Mediterranean Sea gets most of its water from the Atlantic Ocean. So Mediterranean water is very salty.

There are many popular tourist **resorts** along the Mediterranean. These include some of the Mediterranean's many islands, such as Corsica, Sardinia, Sicily, Malta, Crete, and Cyprus. Tourists often like to take a **cruise** across the Mediterranean. They get to visit many different countries all at once, try lots of different food, and see the **remains** of various ancient civilizations.

LEARN MORE! READ THESE ARTICLES...
ATLANTIC OCEAN · ISLANDS · NILE RIVER

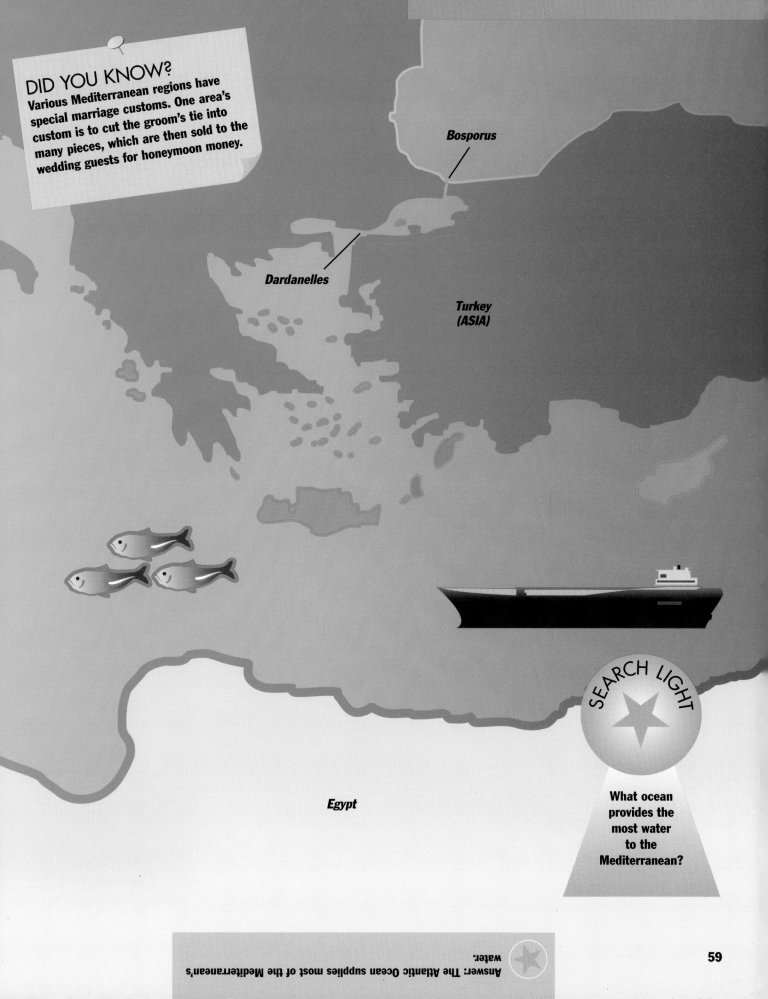

DID YOU KNOW?
Various Mediterranean regions have special marriage customs. One area's custom is to cut the groom's tie into many pieces, which are then sold to the wedding guests for honeymoon money.

Bosporus

Dardanelles

Turkey
(ASIA)

Egypt

SEARCH LIGHT

What ocean provides the most water to the Mediterranean?

Answer: The Atlantic Ocean supplies most of the Mediterranean's water.

DID YOU KNOW?
Whales, the largest animals in the Pacific, have had a strange evolutionary journey. Their early ancestors were land mammals with legs but eventually came to live in the sea and became whales.

AUSTRALIA

SEARCH LIGHT

Fill in the blanks: The Pacific is the

_____ and the

_____ of all the oceans.

Largest Ocean in the World

NORTH AMERICA

The Pacific Ocean is the largest ocean in the world. It covers nearly one-third of the Earth. The Pacific is also deeper than any other ocean. The Pacific Ocean lies between the **continents** of Asia and Australia on the west and North and South America on the east.

The Pacific's deepest parts are the ocean trenches. These trenches are long, narrow, steep, and very deep holes at the bottom of the ocean. Of the 20 major trenches in the world, 17 are in the Pacific Ocean. The deepest trench is the Mariana Trench. It is deeper than Mount Everest, the highest mountain on land, is tall.

There are also many islands in the Pacific Ocean. Some islands were once part of the continents. Some that were part of Asia and Australia include Taiwan, the Philippines, Indonesia, Japan, and New Zealand.

Other Pacific islands have risen up from the floor of the ocean. Many of them are born from volcanoes. These islands are built over thousands of years by the lava that comes out of the volcanoes. The Hawaiian Islands and the Galapagos, for example, started as volcanoes.

The Pacific Ocean is very rich in **minerals**. It also has large supplies of oil and natural gas. And there is rich **marine** life in the Pacific. Fish such as salmon in northwestern America, bonito and prawns in Japan and Russia, and anchovy in Peru are all major food sources for people worldwide.

LEARN MORE! READ THESE ARTICLES...
ATLANTIC OCEAN • GALAPAGOS ISLANDS • ISLANDS

SOUTH AMERICA

Answer: The Pacific is the largest and the deepest of all the oceans.

G L O S S A R Y

alpine relating to mountainous or hilly areas above the line where trees grow

barrier object that blocks access to another object or place; also, something that prevents something else from happening

breadth width

buff an off-white color

canopy overhead covering

channel lengthwise waterway that connects with other bodies of water

circulate flow

clockwise in the direction that a clock's hands move, as viewed from the front

cloudburst sudden heavy rainfall

continent one of the largest of Earth's landmasses

counterclockwise in the direction opposite to the way a clock's hands move, as viewed from the front

crude oil oil taken from the ground and not yet separated into different products; also called petroleum

cruise a pleasure trip on a large boat or ship

debris trash or fragments

drastic huge or dramatic

ecosystem community of all the living things in a region, their physical environment, and all their interrelationships

erode wear down

evaporate change into a vapor or gaseous form, usually by means of heating

exotic unusual and unfamiliar

frigid frozen or extremely cold

gorge narrow steep-walled canyon

gravity force that attracts objects to each other, keeps people and objects anchored to the ground, and keeps planets circling the Sun

habitat the physical environment in which a living thing dwells

handiwork creative product

harness control, much as an animal may be hitched up and controlled by its harness

hemisphere half of the planet Earth or any other globe-shaped object

marine having to do with the ocean

mineral substance that is not animal or plant, and is an important nutrient for living things

mischievous playfully naughty

molten melted

network complex system

overwhelm defeat, beat down, or swallow up

particle tiny bit or piece

peninsula a finger of land with water on three sides

pesticide poison that kills insects dangerous to growing plants

political something related to politics or government

recycle to pass used or useless material through various changes in order to create new useful products from it

remains (noun) parts that are left after time passes or some event occurs

resort (noun) fancy vacation spot

rotate (noun: rotation) spin or turn

sanctuary safe place

species group of living things that have certain characteristics in common and share a name

teeming crowded

timber wood that is cut down for use in building something